THE MATH TUTOR'S FRIEND

Laplace transforms

derivatives

inflection point

BOOK 8

integrals

extrema

tangent

A LOOK AT CALCULUS

by
Joe Cyr

Cyr, Joseph
The Math Tutor's Friend / Book 8 – A Look at Calculus / by Joe Cyr
Third Edition ISBN: 978-1-967039-49-4 Joseph Cyr
Second Edition ISBN: 979-8-9879759-8-5 Joseph Cyr
First Edition ISBN: 9798481056296 Amazon Kindle, 2022

3 4 5 6 7 8 9

Subjects: Main category of book – *introduction to calculus, derivatives, integrals, Laplace Transforms* – high-school math
2.Other category - math tutor reference

Cover illustrations by Ramon E. Owen. Cover Design by Joe Cyr

Book design by Joe Cyr

The Math Tutor's Friend
Series

Books

1 - Pre-algebra
2 - Algebra I
3 - Algebra II
4 - Algebra III
5 - Probability & Statistics
6 - Geometry
7 - Trigonometry
8 - Pre-calculus
9 - Series index & "FLASHMATH" handouts/reference cards

THE MATH TUTOR'S FRIEND

BOOK 8

A Look at Calculus

Sections

1. Derivatives

2. Integrals

3. Laplace Transforms

I Index of Terms

Q Test Questions

About the "Math Tutor's Friend" Series:
Teachers impart knowledge guided by lesson plans in a course syllabus. Math tutors, on the other hand, face varying topics at differing levels in successive sessions; for example, pre-algebra, advanced algebra, geometry, trigonometry, statistics, or even calculus! This series addresses the issue and is intended to provide the tutor with needed quick review of high-school math topics. The material is not presented rigorously nor does it abound with proofs and explanation that are available in textbooks - the tutor already knows this stuff - but a quick refresher is always a good move before diving into a tutoring session.
Too many in high-school have reached the 11th and 12th grades lacking basic math skills such as knowing the multiplication table (relying on a calculator to multiply 6 by 7, for example), handling signed numbers, and performing operations involving fractions - knowledge that should have been acquired by the fifth grade! I try to address this deficiency by devoting the first few tutoring sessions to these fundamentals and providing handouts of multiplication tables,etc.

Books 1 through 8 of this series include a number of quiz questions (answers explained) that the math tutor may find useful.
Book 9 is an overall index of topics. It contains a set of summary "FLASHMATH" pages for quick reference or for use as handouts.
This third edition corrects typos found in previous editions and expands discussion of some of the topics.
Happy Tutoring!

Joe Cyr
Sarasota, FL

●

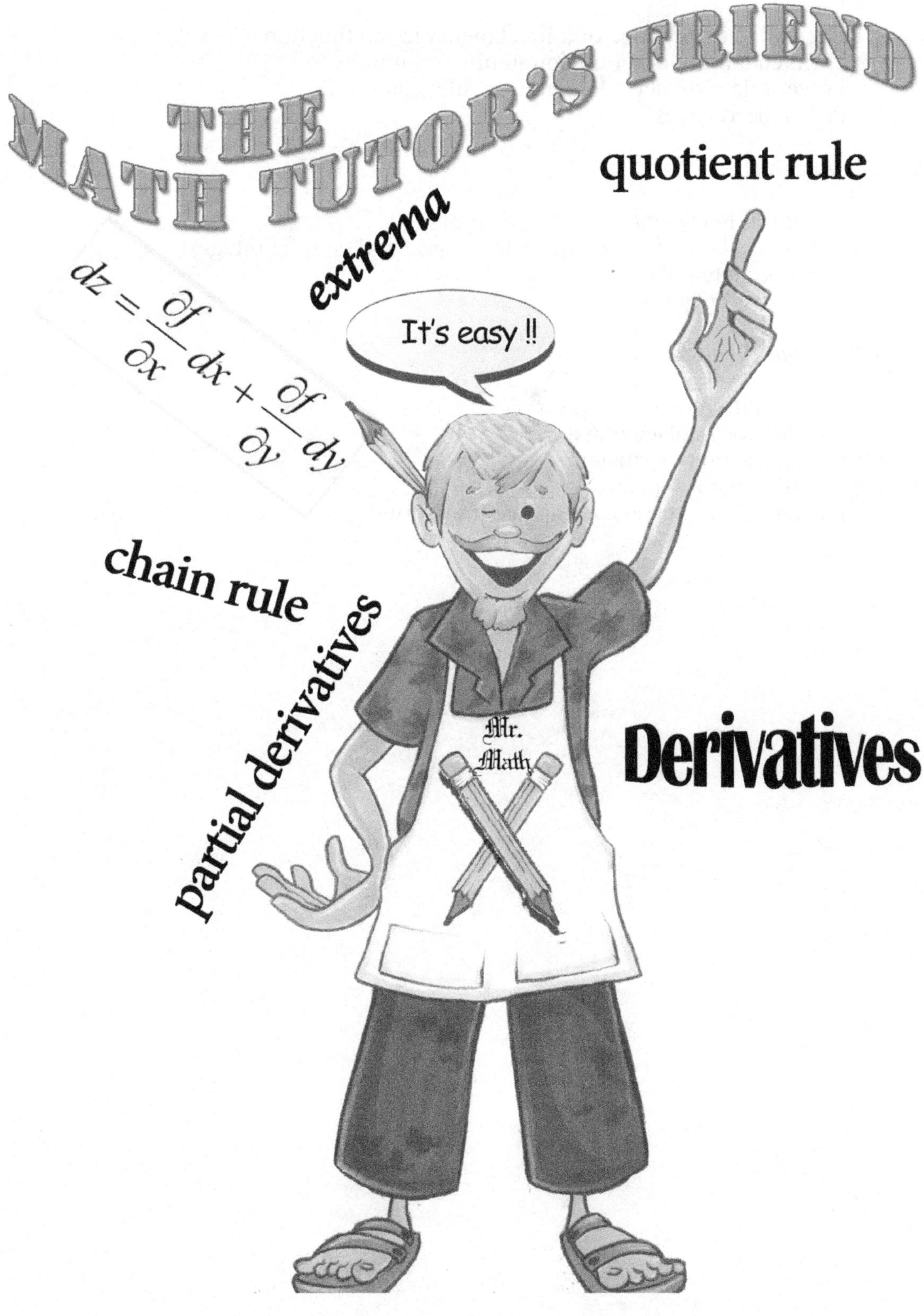
THE MATH TUTOR'S FRIEND
quotient rule
extrema
$dz = \frac{\partial f}{\partial x}dx + \frac{\partial f}{\partial y}dy$
It's easy !!
chain rule
partial derivatives
Mr. Math
Derivatives

About Book 8: "A Look at Calculus"
Section 1, *Derivatives.*
Emphasis -:

- Limits of a function
- Definition of a derivative
- Derivative is the slope of a line tangent to the function
- Differentiability of a function implies continuity
- Power rule; product rule; quotient rule; chain rule
- Partial derivatives

Section 2, *Integrals.*
Emphasis -:

- Graphical interpretation of an integral
- Difference between the indefinite integral and definite integral
- Improper integrals
- Integration by parts

Section 3, *Laplace Transforms.*
Emphasis -:

- Integro-differential equations
- Purpose for Laplace transforms
- Partial fraction expansion
- Use of transform pairs
- Indroduction to tables of Laplace transforms

• •

1. DERIVATIVES
TABLE OF CONTENTS

page

Limit of a function 1-1

Derivatives 1-2
Differentiability and derivative of a function
Differentiability
What is the derivative?

Definition of a derivative 1-3

The Power Rule. 1-5
Comments on notation

Rules for derivatives 1-6
Derivatives of constants, coefficients, and powers of x.
Product and Quotient rules
Product rule (derivative of products)

Chain Rule 1-7

Quotient rule 1-8

Derivatives of other functions. 1-9
Derivatives of trigonometric functions

Derivatives of inverse trigonometric functions. 1-10
Derivatives of complex variables
Example problems involving derivatives

Partial derivatives 1-11

Second-and-higher partial derivatives 1-12
Second partial derivatives
Mixed partial derivatives
Implicit differentiation

•

1. DERIVATIVES

Limit of a function

What is a limit in math? The *limit of a function* f(x) as x approaches p is a number $\boldsymbol{L}$ such that, given any target distance from $\boldsymbol{L}$, there is a distance from p within which the values of f(x) remain within the target distance. That is to say that one can make $f(x)$ as close to $\boldsymbol{L}$ as desired, by making x close enough, but not equal, to p:

$$\lim_{x \to \infty} \frac{1}{x} = 0.$$

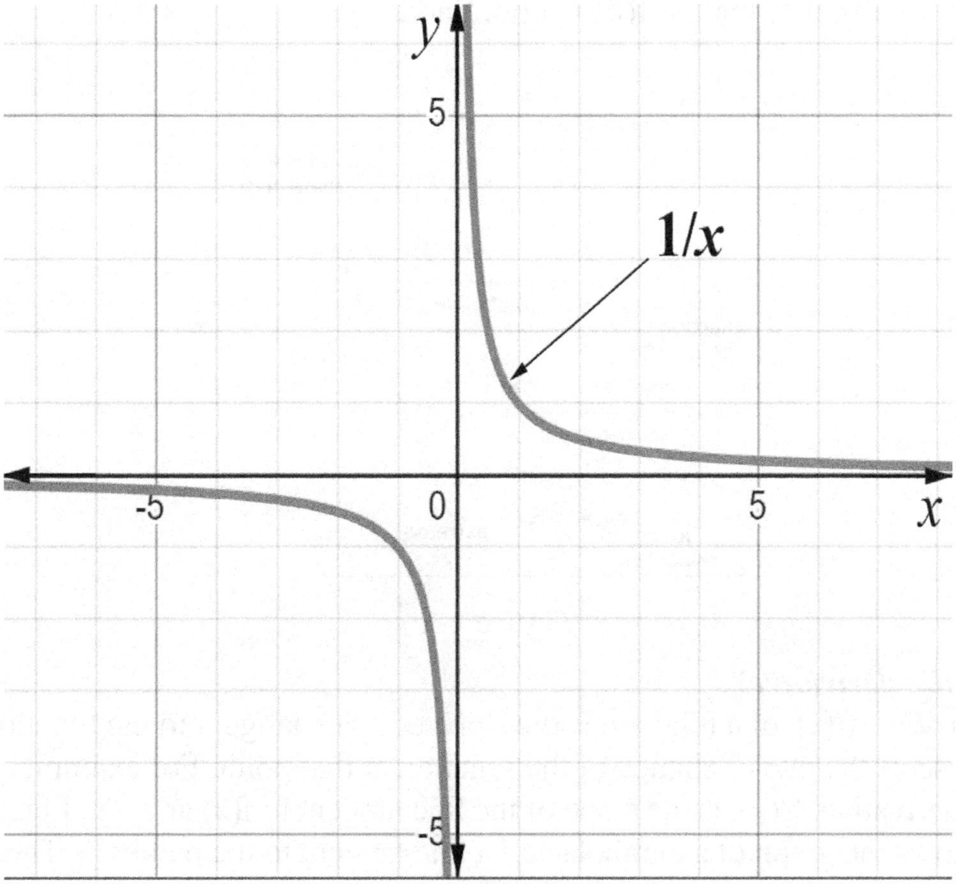

Figure 1-1

Example: The function in Figure 1-1, f(x) = 1/x, has domain $(-\infty, 0) \cup (0, \infty)$ and has

$$\lim_{x \to p} \mathrm{f}(x) = \boldsymbol{L}$$

The function $1/x$ has a horizontal asymptote (x-axis) at $y = 0$ and a vertical asymptote (y-axis) at $x = 0$. All this means is that f(x) will get as close to zero as you wish by making x as large as you like (horizontal asymptote), but $1/x$ can never equal zero. Also as x approaches 0, $1/x$ becomes larger and larger, but x cannot reach zero (vertical asymptote).

Derivatives

Differentiability and derivative of a function

Differentiability

A *differentiable function* is a function whose derivative exists at each point in its domain. The graph of a differentiable function has a non-vertical tangent line at each interior point in its domain. A differentiable function is smooth and contains no break, hole, nor vertical asymptote. Refer to Figure 1-2 below which shows a function that is differentiable at all values of x except at $x = 0$, where a "hole" exist; that is, f(0) is undefined.

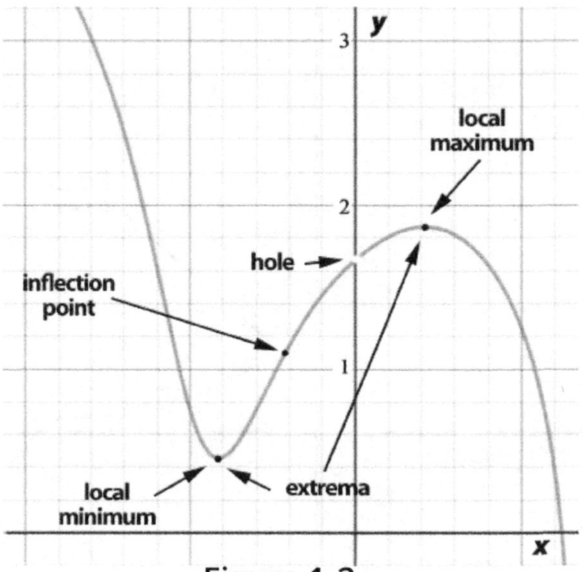

Figure 1-2

What is the derivative?

The *derivative,* $f'(x)$, of a function is the slope of a line tangent to the function at x and represents the rate of change of the function at that point. For example, given $f(x)$, the derivative, $f'(a)$, is the slope of the line tangent to $f(x)$ at $x = a$. Figure 1-3 below shows the graph of a parabola and a line tangent to the parabola. The slope

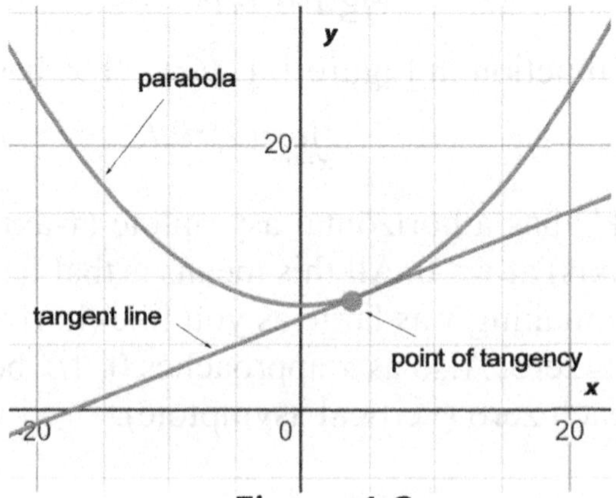

Figure 1-3

of that line is the value of the derivative evaluated at $x = a$. Imagine a different point a small distance to the right of the one shown in Figure 1-3. Clearly the slope of a tangent to the parabola at the new point will be greater; that is, the tangent line will be "steeper." Now envision another point on the parabola at $x = 0$ (that is, to the left of the point shown in Figure 1-3). Surely, the slope of a line tangent to the curve at that point will be zero, i.e., a horizontal line. Further, as the point slides to the left, the tangent line will have a negative slope that becomes steeper and steeper as we move along the curve on the left. Be sure that you understand this intuitive analysis as we re-direct our attention to Figure 1-2 on the preceding page.

Begin by visualizing a point on the function at the top left of the graph. Clearly, the line tangent to that point will have a large negative slope. As we move the point toward the point labeled "local minimum," the slope of the tangent line will remain negative and as it nears the local minimum the slope will become less and less negative until it becomes zero at the local minimum. Continuing to the right, the slope of the tangent line is positive and will increase in magnitude until it reaches the "inflection point" at which time it will become less and less positive until it reaches $x = 0$ where $f(x)$ is undefined; moving past $x = 0$ the slope of the tangent line becomes less and less positive until it reaches zero at the point labeled "local maximum," after which the slope becomes increasingly negative as we move on to lower right of the graph. The value of the slope of a line tangent to any x in the domain of the graph is the derivative of the function evaluated at that value of x.

If the *first derivative* of a function is itself a continuous function within some domain, one may take second (and higher) derivatives of that function within the domain. For example, the first derivative, $f'(x)$, of the function

$$f(x) = 3x^3 - 2x^2 + 5x - 3 \text{ is } f'(x) = 9x^2 - 4x + 5,$$

which itself is differentiable. The second derivative, $f''(x) = 18x - 4$, describes the rate of change of the slope at any point. For example, we saw that the slope changed at the inflection point from increasingly positive to decreasingly positive. The *second derivative* of a function, $f''(x) = f'(f'(x))$, changes sign at an *inflection point* and so becomes zero at that inflection point, which is where the function switches from being *convex* to being *concave* or vice versa.

For *extrema*, $f''(x) < 0$ and $f''(x) > 0$ indicate a maximum and a minimum, respectively.

Definition of a derivative

The *derivative* of $f(x)$ with respect to x *at* $x = a$, denoted by $f'(a)$, is defined as

$$f'(a) = \lim_{h \to 0} \frac{f(a+h) - f(a)}{h}$$

Let's look at the graph of f(x) in Figure 1-4.

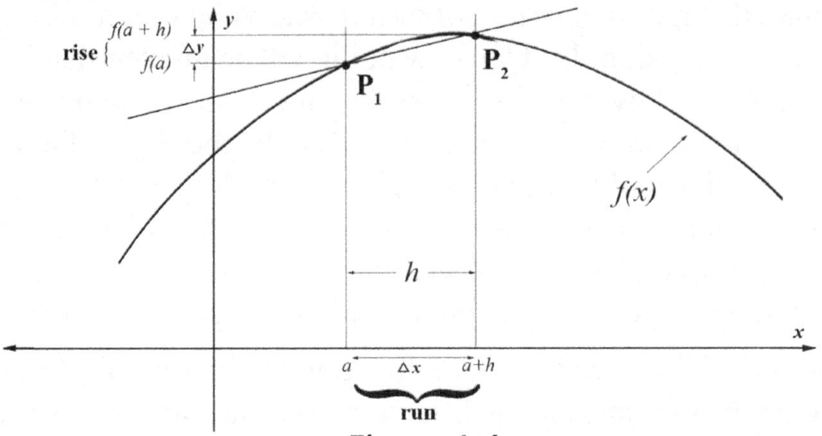

Figure 1-4

Select a value $x = a$; point $\mathbf{P}_1$ is f(a). Select another value, $x + h$, a distance, h, from a; f($a + h$) is point $\mathbf{P}_2$. Draw a secant line that passes through P_1 and P_2 as shown. The slope of the secant line is the "rise over the run," that is,

$$\frac{f(a+h)-f(a)}{h}$$

As the "run" gets smaller, and approaches zero, the distance between points P_1 and P_2 approaches zero also. In the "limit" the points merge and the secant line is tangent to the curve at that point. This is all summarized as follows:

$$f'(a) = \lim_{h\to 0}\frac{f(a+h)-f(a)}{h} = \frac{y_2 - y_1}{x_2 - x_1} = \frac{\Delta y}{\Delta x} = \frac{rise}{run} = m = slope$$

The "h" in the formula really means a "change in x" sometimes expressed as "Δx" as shown above.

Example: Use the definition to find the derivative of

$$f(x) = 3x^2 - 5x + 4$$

Applying the definition of a derivative we have

$$f'(x) = \lim_{h\to 0}\frac{3(x+h)^2 - 5(x+h) + 4 - (3x^2 - 5x + 4)}{h}$$

Expanding the binomial

$$= \lim_{h\to 0}\frac{3(x^2 + 2hx + h^2) - 5(x+h) + 4 - (3x^2 - 5x + 4)}{h}$$

Removing parentheses

$$= \lim_{h\to 0}\frac{3x^2 + 6hx + 3h^2 - 5x - 5h + 4 - 3x^2 + 5x - 4}{h}$$

Simplifying the numerator ("like" terms add to zero):

$$= \lim_{h \to 0} \frac{h(6x + 3h - 5)}{h}$$

which, after cancelling the *h*'s, becomes

$$= \lim_{h \to 0} (6x + 3h - 5)$$

As *h* approaches zero so does 3*h*.

Thus, for $f(x) = 3x^2 - 5x + 4$ we have $f'(x) = 6x - 5$.

The Power Rule

There is usually no need to employ the definition to find a derivative. Note that in the example the first term of $f'(x)$ is the product of the exponent of the first term of $f(x)$ and its coefficient, i.e., $2 \times 3 = 6$ with the exponent reduced by 1 (i.e., x^2 becomes $x^1 = x$). The same "rule" applies to all of the terms of $f(x)$, so $5x$ becomes $5x^1 = 1 \times 5x^0 = 5$ and the constant "4" is really $4x^0$ and applying the "rule" we have $0 \times 5x^{-1}$. Put all of this together and we see that the derivative of $f(x) = 3x^2 - 5x^1 + 4x^0$ is

$$f'(x) = 2 \times 3x^{2-1} - 1 \times 5x^{1-1} + 0 \times 4^{0-1} = 6x - 5$$

as already determined by applying the definition of the derivative. The above procedure is an application of the *Power Rule* for differentiation. There are comparable rules for other types of functions such as trig functions, log functions, etc. and the products and other combinations of these. Tables of derivatives can be found in all calculus texts and on the internet.

Comments on notation

Before proceeding let's look at some of the notation used to indicate the derivative:

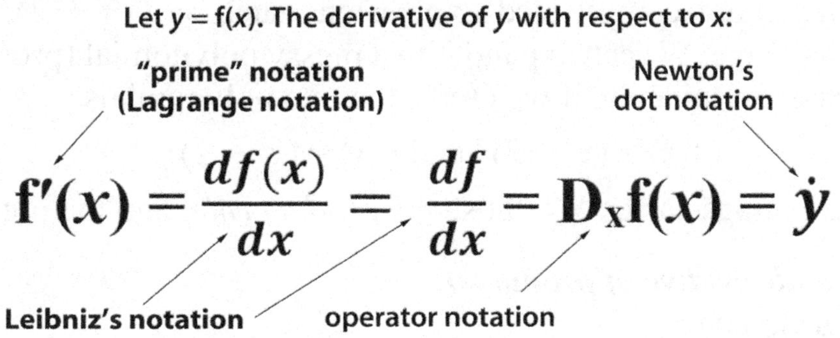

In the previous example we used $f'(x)$, called the *prime* or *Lagrange notation*. The second, third and fourth notations are more-or-less the same where $f(x)$ is expressed as " f " or " *y*. " So we call the derivative of *y* with respect to *x* "dee eff dee *x*" or "dee *y* dee *x*." The *Leibnitz notation* is probably the most common notation. Next, $D_x f(x)$, the *operator notation*, is read as "D sub *x* of eff of *x*." The last notation is *Newton's dot notation*.

It is possible to take a derivative of a derivative, and so on. The notation for a "second derivative" is " f'' " or " d^2y/dx^2. " The notations dy/dx and d^2y/dx^2 represent the derivative of y with respect to x and the second derivative of y with respect to x squared, respectively.

There is another notation, $\partial y/\partial x$, indicating *partial differentiation.* For example we may want to differentiate a function of, say, three variables such as $f(x, y, z)$ with respect only to x and so, the notation is $\partial f/\partial x$; the variables y and z are treated as constants.

Rules for derivatives

(*Rules of Differentiation*)

The procedure discussed so far works for the standard polynomial. We'll continue with brief descriptions about handling derivatives of constants and other types of functions.

Derivatives of constants, coefficients, and powers of *x*.

The last is the Power Rule already discussed.

Let $f(x)$ be a differentiable function and let c be a constant; then

$$\frac{d}{dx}c = 0 \text{ [the derivative of a constant is zero]}$$

$$\frac{d}{dx}(cx - c) = \frac{d}{dx}cx - \frac{d}{dx}c = c - 0 = c$$

$$\text{Let } f(x) = x^n \text{ then } f'(x) = \frac{d}{dx}x^n = nx^{n-1}$$

Product and Quotient rules

Differentiating more complicated expressions such as $(x^2 + 2)^3(x^3 + 1)^2$ presents a challenge. We can expand it to a messy polynomial (prone to errors in the process). Fortunately, if we create the two polynomials

$$f(x) = (x^2 + 2)^3 \text{ and } g(x) = (x^3 + 1)^2,$$

it might make things easier. We'll use the *product rule*, and the notation $\frac{dy}{dx}$.

Product rule (derivative of products)

Let $y = f(x)\cdot g(x)$, then

$$\frac{dy}{dx} = f'(x)\cdot g(x) + f(x)\cdot g'(x)$$

Proof:

Use the definition to obtain the rule of differentiation for the product of functions. Let u and v be differentiable functions of x; that is, $u = u(x)$ and $v = v(x)$, and let $y = u\cdot v$ and $h = \Delta x$.

1) Apply the definition of a derivative:

$$\frac{d(u\cdot v)}{dx}=\lim_{h\to 0}\frac{u(x+h)v(x+h)-u(x)v(x)}{h}$$

2) Add $u(x+h)v(x)$ and its opposite $-u(x+h)v(x)$ to the numerator (the "trick" of adding zero):

$$\frac{d(u\cdot v)}{dx}=\lim_{h\to 0}\frac{[u(x+h)v(x+h)-u(x+h)v(x)]+[u(x+h)v(x)-u(x)v(x)]}{h}$$

3) Rearrange the terms of the numerator into factorable groups then factor out the common factor [$u(x+h)$ and $v(x)$] from each group.

$$\frac{d(u\cdot v)}{dx}=\lim_{h\to 0}\frac{u(x+h)[v(x+h)-v(x)]+v(x)[u(x+h)-u(x)]}{h}$$

4) Rearrange, noting that $\lim_{h\to 0}\frac{y\cdot z}{h}=\lim_{h\to 0}y\cdot\lim_{h\to 0}\frac{z}{h}$

and $\lim_{h\to 0}v(x)=v(x)$

$$\frac{d(u\cdot v)}{dx}=\lim_{h\to 0}u(x+h)\cdot\lim_{h\to 0}\frac{[v(x+h)-v(x)]}{h}+v(x)\cdot\lim_{h\to 0}\frac{[u(x+h)-u(x)]}{h}$$

Note also that . $\lim_{h\to 0}u(x+h)=u(x)$

$$\frac{d(u\cdot v)}{dx}=u(x)\cdot\lim_{h\to 0}\frac{[v(x+h)-v(x)]}{h}+v(x)\cdot\lim_{h\to 0}\frac{[u(x+h)-u(x)]}{h}$$

6) Apply the definition of the derivative to v and u:

$$\lim_{h\to 0}\frac{f(x+h)-f(x)}{h}=\frac{df}{dx}$$

$$\frac{d(u\cdot v)}{dx}=u(x)\cdot\frac{dv}{dx}+v(x)\cdot\frac{du}{dx}$$

This concludes the proof for the derivative of the product f(x)g(x):

$$d(f(x)g(x))/dx = f'(x)g(x)+f(x)g'(x)\ .$$

Chain Rule

What is the derivative of the composition of two functions? Recall that f(g(x)), also expressed as "f∘g", is the *composition of the two functions* f(x) and g(x).

Here's the *chain rule* for f(x) and g(x) showing the two equivalent forms:

$$(f \circ g)'(x) = f'(g(x))g'(x)$$

Example: Consider the function

$$f(x) = (x + 2)^2 \text{ with its derivative } f'(x) = 2(x + 2)$$

Let $f(x) = [g(x)]^2$, where $g(x) = (x + 2)$ and $g'(x) = 1$.

Then f can be expressed as the composite function f(g(x))

Using the chain rule: $f = f(g(x))$ and $\frac{df}{dx} = \frac{df}{dg} \cdot \frac{dg}{dx}$

$f'(x) = f'(g(x))g'(x) = 2(x + 2)\cdot 1 = 2(x + 2)$

Example:

Differentiate the following: $y = x^3(4x^2 + 1)^5$

Let $\boldsymbol{u} = x^3$ then $\boldsymbol{u}' = 3x^2$

$\boldsymbol{v} = (4x^2 + 1)^5$ then $\boldsymbol{v}' = 5(4x^2 + 1)^4 \cdot 8x = 40x(4x^2 + 1)^4$

$$y' = \boldsymbol{u}'\,\boldsymbol{v} + \boldsymbol{v}'\,\boldsymbol{u}$$

$$y' = 3x^2(4x^2 + 1)^5 + 40x(4x^2 + 1)^4 x^3$$

$$y' = 3x^2(4x^2 + 1)^5 + 40x^4(4x^2 + 1)^4$$

factor out $x^2(4x^2 + 1)^4$:

$$y' = x^2(4x^2 + 1)^4[3(4x^2 + 1) + 40x^2]$$

Remove parentheses and combine terms within the brackets [ ] :

$$y' = x^2(4x^2 + 1)^4(12x^2 + 3 + 40x^2)$$

$$y' = x^2(4x^2 + 1)^4(52x^2 + 3)$$

Note that both the product and chain rules were used.

Quotient rule

The *quotient rule* is a special case of the product rule because the rational function f(x)/g(x) can be written as the product $f(x)\cdot[g(x)]^{-1}$.

Let h(x) be a rational function: $h(x) = \frac{f(x)}{g(x)}$

Then we can write f(x) = h(x)g(x) and from the product rule,

$$f'(x) = g'(x)h(x) + g(x)h'(x).$$

Solve the above for h′(x):

$$g(x)h'(x) = f'(x) - g'(x)h(x)$$

$$h'(x) = \frac{f'(x) - g'(x)h(x)}{g(x)}$$

but

$$h(x) = \frac{f(x)}{g(x)}$$

so,

$$h'(x) = \frac{f'(x) - g'(x)\frac{f(x)}{g(x)}}{g(x)}$$

$$h'(x) = \frac{f'(x)g(x) - g'(x)f(x)}{[g(x)]^2}$$

which is the quotient rule; that is,

$$\frac{d}{dx}\frac{f(x)}{g(x)} = \frac{f'(x)g(x) - f(x)g'(x)}{[g(x)]^2}$$

More concisely, consider the two functions $u(x)$ and $v(x)$. Then the product and quotient rules can be written as

$$d(uv) = u'v + v'u$$

and

$$\frac{d}{dx}\frac{u}{v} = \frac{u'v - v'u}{v^2}$$

Derivatives of other functions

Here is a list of derivatives for several other functions. Tables of derivatives are available in calculus textbooks and on the internet.

$$\frac{d}{dx}e^x = e^x$$

$$\frac{d}{dx}\ln(x) = \frac{1}{x}$$

$$\frac{d}{dx}a^x = a^x \ln(a)$$

$$\frac{d}{dx}\log_a x = \frac{1}{x} \cdot \frac{1}{\ln(a)}$$

Derivatives of trigonometric functions

$$\frac{d}{dx}\tan x = sec^2 x$$

$$\frac{d \sin x}{dx} = \cos x \qquad \frac{d}{dx}\cot x = -csc^2 x$$

$$\frac{d \cos x}{dx} = -\sin x \qquad \frac{d}{dx}\sec x = \sec x \tan x$$

$$\frac{d}{dx}\csc x = -\csc x \cot x$$

Example (*Chain-sine Rule*): Find the derivative of

$$y = 2\sin(2x^2 + 3x)$$

Let u = $2x^2 + 3x$

$$\frac{dy}{dx} = 2\cos u \cdot \frac{du}{dx} = 2\cos(2x^2 + 3x) \cdot (4x + 3)$$

Derivatives of inverse trigonometric functions

$$\frac{d}{dx}\arcsin x = \frac{1}{\sqrt{1 - x^2}}$$

$$\frac{d}{dx}\arccos x = \frac{-1}{\sqrt{1 - x^2}}$$

$$\frac{d}{dx}\arctan x = \frac{1}{1 + x^2}$$

$$\frac{d}{dx}\operatorname{arccot} x = \frac{-1}{1 + x^2}$$

$$\frac{d}{dx}\operatorname{arcsec} x = \frac{1}{|x|\sqrt{x^2 - 1}}$$

$$\frac{d}{dx}\operatorname{arccsc} x = \frac{-1}{|x|\sqrt{x^2 - 1}}$$

Derivatives of complex variables

This is a topic in the mathematical area of complex analysis and is beyond the scope of this book.

Example problems involving derivatives

Example: A ball is dropped from a height of 80 feet. Its height above ground (in feet) t seconds later is given by

$$s(t) = -16t^2 + 80.$$

What is the velocity of the ball when it hits the ground?

Solution:

The height of the ball is zero when it hits the ground;

$$s(t) = -16t^2 + 80 = 0$$

The $-16t^2$ term is the acceleration due to gravity (negative because gravity pulls downward) and the second term, the constant 80, is the position of the ball at time t = 0.

Now solve for t:

$$t^2 = 80/16 \text{ and } t = (\sqrt{80})/4 = \sqrt{(16 \times 5)}/4 = \sqrt{5} \approx 2.2 \text{ seconds}$$

The velocity is the first derivative of s(t):

$$s'(t) = -32t \text{ ft/sec}$$

and so at $t = \sqrt{5}$ seconds the velocity will be

$$-32 \times \sqrt{5} = -71.6 \text{ ft/sec},$$

again, negative because the movement is downward.

What is the acceleration of the ball when it hits the ground? The acceleration is the second derivative of s(t):

$$s''(t) = s'(s'(t)) = -32 \text{ ft/sec}^2,$$

as expected because this is the acceleration due only to gravity (i.e., there was no initial acceleration by, say, throwing the ball down instead of letting it fall). The gravitational constant for Earth is -32.2 ft/sec^2, approximately, and varies slightly depending upon the location (lower at the Equator because of centrifugal force due to the Earth's rotation).

Example: A rectangle has length three times its width. What is the maximum area that can be enclosed by a perimeter of 250 ft?

Solution: Let y and x be length and width.
Given: $y = 3y$ and $2x + 2y = 250$ from which $y = 125 - x$
$A = \text{area} = xy = x\cdot(125 - x) = 125x - x^2$
$A' = 125 - 2x = 0$
at an extremum where $x = 62.5$
$A'' = -2$, so the extremum is a maximum.
The dimensions for maximum area:

$$x = 62.5 \text{ ft and } y = 3x = 3\times 62.5 = 187.5 \text{ ft}$$

and the area is $62.5\times 187.5 \approx 11\,719$ sq ft, a little more than a quarter of an acre.

Partial derivatives

The *total derivative* of a function f(x, y) is given by

$$df = \frac{\partial f}{\partial x}dx + \frac{\partial f}{\partial y}dy = f_x + f_y$$

Example: A cylinder has both radius, r, and height, h, equal to 3 meters. If the radius decreases at a rate of 1 m/min and the height increases at 2 m/min, what is the rate of change of the volume of the cylinder?
The formula for the volume, V, of a cylinder in $V = \pi r^2 h$.

Solution: The rate of volume change is the *total differential*

$$\frac{dV}{dt} = \frac{\partial V}{\partial r}\frac{dr}{dt} + \frac{\partial V}{\partial h}\frac{dh}{dt}$$

Here, $\frac{dr}{dt} = -1 \text{ m/min}$ and $\frac{dh}{dt} = 2 \text{ m/min}$

$$\frac{dV}{dt} = \frac{\partial V}{\partial r}\frac{dr}{dt} + \frac{\partial V}{\partial h}\frac{dh}{dt} = 2\pi rh\times(-1)+\pi r^2$$

$$= -2\pi 3\cdot 3\cdot 1 = 2\pi 3^2 = -18\pi + 18\pi = 0$$

The volume of the cylinder remains unchanged.

Second-and-higher partial derivatives

As with derivatives, there are second (and higher) partial derivatives. For two variables x and y there are four cases:

Second partial derivatives

$\frac{\partial^2 f}{\partial x^2}$ or f_{xx} or $(f_x)_x$ differentiate f with respect to x, then the result with x again.

$\frac{\partial^2 f}{\partial y^2}$ or f_{yy} or $(f_y)_y$ differentiate f with respect to y, then the result with y again.

Mixed partial derivatives

$\frac{\partial^2 f}{\partial x \partial y}$ or f_{xy} or $(f_x)_y$ differentiate f with respect to x, then the result with respect to y.

$\frac{\partial^2 f}{\partial y \partial x}$ or f_{yx} or $(f_y)_x$ differentiate f with respect to y, then the result with respect to x.

The same result usually occurs whatever order the derivatives are taken in mixed partials.

Example: Let $f(x, y) = 4x^2y + 3x - 3y^2 + 2$.
Find f_x, f_y, f_{xx}, f_{yy}, f_{xy}, and f_{yx}. $f_x = 8xy + 3$, $f_y = 4x^2 - xy$,
$f_{xx} = 8y$, $f_{yy} = -6$, $f_{xy} = 8x$, $f_{yx} = 8x$.

Implicit differentiation

Some functions are in explicit form, that is, $y = f(x)$. You might encounter a function that contains a mixture of different variables. It is not always easy to separate the variables in the equation. An example is $y^2 + 2yx + 3 = 5x$. This function is too cumbersome to isolate y. Therefore, one can utilize implicit differentiation to find the derivative. Consider y to be a nested function that is defined implicitly by x and employ the Chain Rule.

For example,

$$\frac{d}{dx}[f(g(x))] = f'(g(x)) \cdot g'(x)$$

or equivalently,

$$\frac{dz}{dx} = \frac{dz}{dy} \cdot \frac{dy}{dx}$$

Example:
Find the derivative of $y^2 + x^2 = 25$ with respect to x.

You are seeking $\frac{dy}{dx}$.

Take the derivative of each side of the equation with respect to x.

$$\frac{d(y^2 + x^2)}{dx} = \frac{d(25)}{dx}$$

$$2y \cdot \frac{dy}{dx} + 2x = 0$$

$$2y \cdot \frac{dy}{dx} = -2x$$

$$\frac{dy}{dx} = -\frac{x}{y}$$

●

THE MATH TUTOR'S FRIEND
Riemann sums
indefinite integral
limits of integration
It's easy !!
y=f(x)
a
b
x
y
Mr. Math
Integrals
integrate by parts

2. INTEGRALS
TABLE OF CONTENTS

page

Integration 2-1
Indefinite integral
Area under a curve

Definite Integral 2-2
Riemann sums

Improper Integrals 2-3
Anatomy of a Definite Integral

Anti-derivatives 2-4

Indefinite Integral 2-5
Anatomy of an Indefinite Integral

List of integrals 2-6

Integrals of inverse trig functions 2-7
Integration by parts

•

2. INTEGRATION

Integration

The *Fundamental Theorem of Calculus* essentially states that differentiation and integration are related. *Integration* is the inverse process of differentiation. A function is *integrable* if the integral over its domain is finite. The two categories of integration are the *indefinite integral* and the *definite integral*, the latter having specified *limits of integration.* The result of calculating an indefinite integral is a function; the result of solving a definite integral is a function evaluated at the upper limit of integration minus the same function evaluated at the lower limit of integration. The definite integral is mainly used to calculate the area under a curve; the indefinite integral determines the *antiderivatives* (f from f′).

Indefinite integral

The indefinite integral is also called the *antiderivative* of a function f(x). Let $dy/dx = x^3$ and suppose we want to find y. We can write $dy = x^3dx$ and then $y = \int x^3\,dx$ is the antiderivative of $dy = x^3dx$. Stated another way, we ask "what is the function f(x) that, when differentiated, will result in x^3?" Well, if we differentiate x^4 we'll get $4x^3$ and we're almost there except for that coefficient of 4, so let's differentiate $(1/4)x^4$ to get $4(1/4)x^3$ which yields the desired result x^3. That makes $(1/4)\,x^4$ the antiderivative of x^3 and we write $\int x^3dx = (1/4)x^4$. That is, if $F'(x) = f(x)$, then $F'(x)$ is a derivative and $F(x)$ its *antiderivative*, so $F(x) = \int f(x)dx$.

<u>*But Wait! Don't forget the constant:*</u> We saw that the derivative of $(1/4)x^4$ is x^3, but also, the derivative of $(1/4)\,x^4 + 12$ is x^3, the derivative of $(1/4)\,x^4 - 2000$ is x^3, and the derivative of $(1/4)x^4 + c$ is x^3, where c is any constant. So, how do we know which constant to insert in the antiderivative? The answer, of course, is "we don't." The unknown c is the *constant of integration*, and we write $y = \int x^3\,dx = (1/4)\,x^4 + c$. Again, note that the indefinite integral yields a function.

Area under a curve

How does one determine the area under the graph of a function within a specified interval on the x-axis? One way to estimate such an area is to divide the area into rectangles and then calculate the sum of the areas of these rectangles. Intuitively, the "thinner" the rectangles, the closer the approximation. This approximation technique is called the *Riemann Sum.* Let's call the width of each rectangle Δx. As Δx becomes smaller the rectangles become more numerous and the approximation of the area (the sum of the

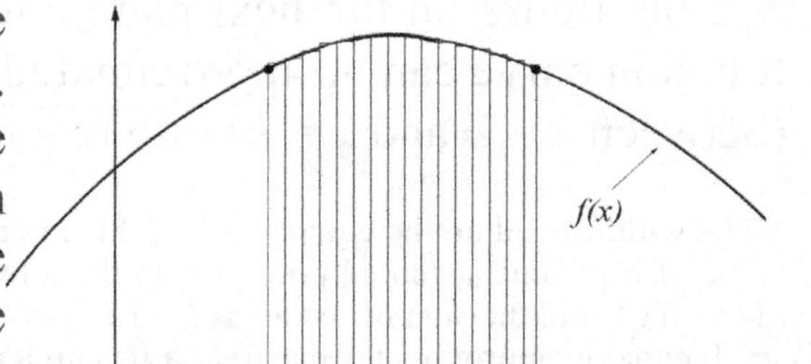

areas of the rectanglers) becomes more accurate. In the limit; that is as $\Delta x \to 0$, the area will be exact. Enter the *definite integral.*

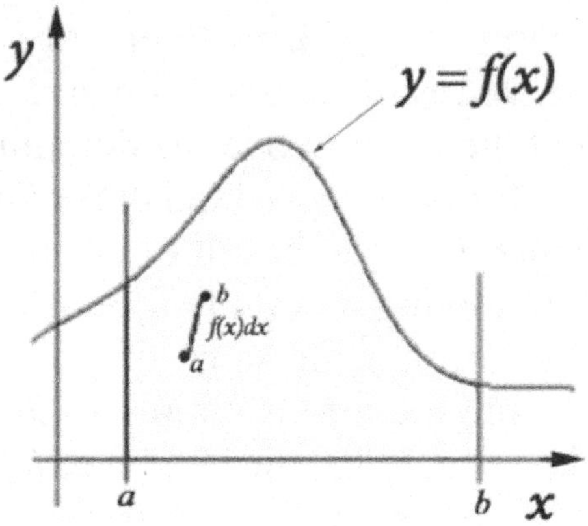

Definite Integral

Consider a function *f* over an interval $[a, b]$ of its domain. The *definite integral* is the area of the region in the *xy* plane bounded by the graph of *f*, the *x*-axis, and vertical lines drawn at $x = a$ and $x = b$. See the figure above. In the case where the function crosses the *x*-axis within the specified interval, the area below the *x*-axis is subtracted from the area above the *x*-axis. The definite integral specifies an upper and lower limit which are the *x*-values of the end-points of the interval $[a, b]$. As already mentioned, the indefinite integral produces a function (and a constant) and the definite integral produces a number, namely, the area under the curve.

The definite integral specifies an upper and lower limit for *x*. See the figure below. The shaded area under the function $y = f(x)$ between $x = a$ and $x = b$ is given by the definite integral.

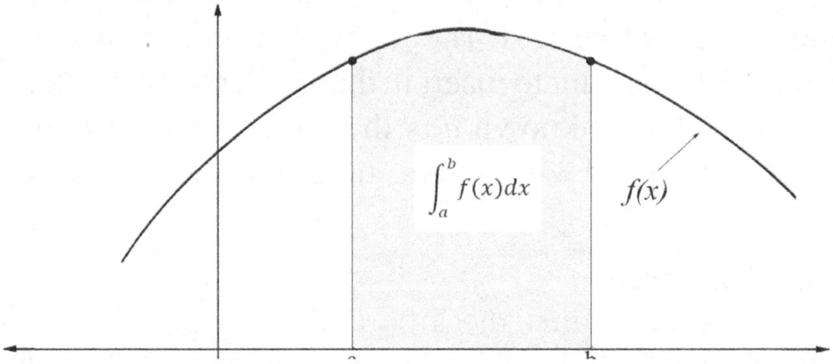

Riemann sums

See the figure on the next page; one can easily see that the area under the function curve can be approximated by partitioning the area under the curve (between $x = a$ and $x = b$) into rectangles of some width Δx as shown*.

* The widths need not be equal; each of the *n* rectangles, *i*, has width Δx_i and length $f(x_i)$ where x_i is the value of *x* at some specified point within the interval (e.g., left, center, right, ...). The area of rectangle *i* is $\Delta x_i f(x_i)$, and the sum of the areas $i = 1$ to $i = n$ is the Riemann sum which, if the rectangle widths are small enough, approximates the area of f(x) under the curve (beneath $x = a$ and $x = b$). Areas beneath the *x*-axis above the curve are subtracted from areas above the axis below the curve.

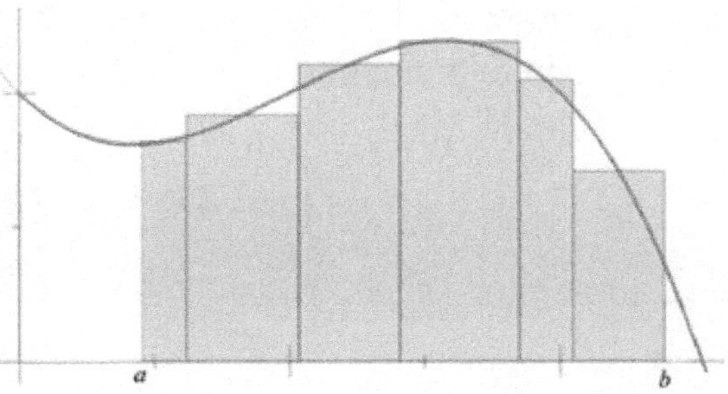

Clearly, very small values of Δx for many rectangles will "fit" the area more closely, and that the sum of all of the areas of the rectangles will more closely estimate the area under the curve between $x = a$ and $x = b$. As the rectangles get smaller and smaller, the sum, called the Riemann sum, approaches the value of the integral between $x = a$ and $x = b$. Other "tricks" could be used to make a closer approximation; for example, the tops of the rectangles could intersect the curve (as shown in the figure), or a top could be slanted to approximate the slope of the segment of the curve between the sides of the shape (which would then be trapezoids). There is much more to Riemann sums. This paragraph only introduces the topic.

Improper Integrals

Definite integrals where one or both of the limits tend to infinity are called *improper integrals*. For example,

$$\int_a^{\infty} f(x) = \lim_{b\to\infty} \int_a^b f(x) = \lim_{b\to\infty} F(b) - F(a).$$

Example:

$$\int_0^{\infty} 3e^{-x} = -3(e^{-\infty} - e^{-0}) = -3(0 - 1) = 3$$

Examples of using the "∞" symbol as limits of integration:

$$\int_{-\infty}^{\infty} f(x)dx \; ; \; \int_{-\infty}^{a} f(x)dx \; ; \; \int_b^{\infty} f(x)dx$$

In general, the integral of a real-valued function f(x) with respect to a real variable x on an interval $[a, b]$ is written as

$$\int_a^b f(x)dx.$$

Anatomy of a Definite Integral

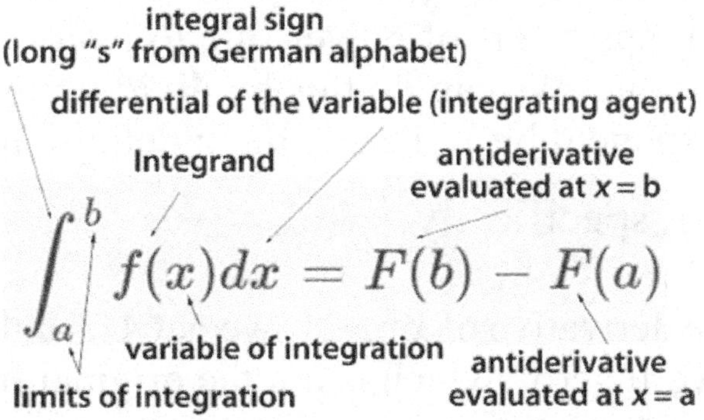

Example: Find the area under the line $y = 2x + 2$ between $x = 1$ and $x = 10$.
Set up the definite integral as shown below:

$$\int_1^{10} (2x+2)dx = \int_1^{10} 2xdx + \int_1^{10} 2dx$$

$$= x^2\Big|_1^{10} + 2x\Big|_1^{10}$$

$$= (100 - 1) + (20 - 2) = 99 + 18 = 117$$

The vertical bars indicate that the answer is to be evaluated at 10 and 1, as shown. What is really happening is that the area under the line from $x = 0$ to $x = 1$ is subtracted from the area from $x = 0$ to $x = 10$.

Anti-derivatives

The indefinite integral, $\int f(x)dx$, represents a class of functions (anti-derivative) whose derivative is the integrand.

Example:

$\int 4x^3 dx = x^4$ or $x^4 + 5$ or $x^4 + 247$ or $x^4 - 45$ or $x^4 + 1$, *ad infinitum*

because the derivative of any of these is identical: $4x^3$. The correct expression is

$$\int 4x^3 dx = x^4 + C, \text{ where C is a constant.}$$

However,

$$\int_2^3 4x^3\, dx = 3^4 - 2^4 = 81 - 16 = 65.$$

Note that the indefinite integral produces a function (which, when differentiated yields the integrand) and the definite integral produces a number.

When we perform the derivative of $\frac{dy}{dx} x^4$ to get $4x^3$ using the power rule we may be interested in the reverse procedure or anti-derivative of $4x^3$ to get the original x^4. Perhaps a sort of power rule to restore a function like $4x^3$ to a 4th power variable. We can do this by dividing the coefficient by 4 and increasing the exponent by 1. This is the indefinite integral

$$\int f(x), \text{ specifically } \int 4x^3 = \frac{4}{4} x^4 = x^4$$

But wait! If we take the derivative of $x^4 + 21$, we get $4x^3$ and if we "reverse" this as shown above, we'll get x^4 which is not the original function $x^4 + 21$.

Indefinite Integral

The integration of a function f(x), i.e., the integral of f(x) with respect to x, is sometimes denoted by F(x):

$$\int f(x)dx = \mathrm{F}(x) + \mathrm{C}$$

where F(x) is the anti-derivative, x is the variable of integration, f(x) is the integrand, dx is the *integrating agent*, and C is the *constant of integration*. Indefinite integrals have neither upper nor lower limits and represent the family of a function whose derivatives are f. That is, there exists an infinite number of anti-derivatives for a function ***f*** because of C, an arbitrary constant that can be any value including zero.

Anatomy of an Indefinite Integral

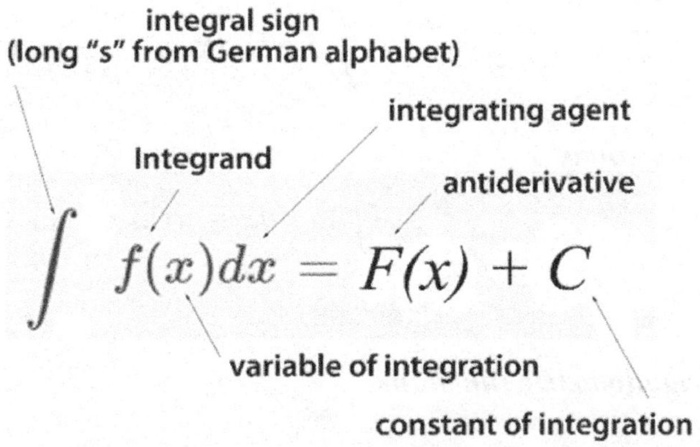

The *Fundamental Theorem of Calculus* relates the evaluation of definite integrals to indefinite integrals. Just as there is a second (and higher) derivatives, there are second (and higher) integrations, for example,

$$\iint_a^b f(x)dx^2,\ \iiint f(x)dx^3$$

Other types of multiple integrals are associated with higher dimensions, e.g., (x, y), (x, y, z). See the list of integrals on the following pages. Note that if you take the derivative of the right-hand side of the equations you get the integrand of the left side.

List of integrals

Constant

$$\int a dx = ax + C$$

Exponents

$$\int x^n dx = \frac{x^{n+1}}{n+1} + C$$

$$\int e^x dx = e^x + C$$

$$\int a^x dx = \frac{1}{\ln a} a^x + C$$

$$\int xe^{ax} dx = \frac{e^{ax}}{a^2}(ax - 1) + C$$

Logarithms

$$\int \ln x \, dx = x(\ln x - 1) + C$$

Trigonometric functions

$$\int \sin x \, dx = -\cos x + C$$

$$\int \cos x \, dx = \sin x + C$$

$$\int \tan x \, dx = -\ln|\cos x| + C$$

$$\int \cot x \, dx = \ln|\sin x| + C$$

$$\int \sec x \, dx = \ln|\sec x + \tan x| + C$$

$$\int \csc x \, dx = -\ln|\csc x + \cot x| + C$$

$$\int sec^2 x dx = \tan x + C$$

$$\int csc^2 x dx = -\cot x + C$$

$$\int \sec x \tan x \, dx = \sec x + C$$

$$\int \csc x \cot x \, dx = -\csc x + C$$

List of integrals (continued)

Reciprocals

$$\int \frac{1}{x} dx = \ln|x| + C$$

$$\int \frac{dx}{\sqrt{a^2 - x^2}} = \arcsin\frac{x}{a} + C$$

$$\int \frac{dx}{a^2 + x^2} = \frac{1}{a}\arctan\frac{x}{a} + C$$

$$\int \frac{dx}{x\sqrt{x^2 - a^2}} = \frac{1}{a}\operatorname{arcsec}\frac{|x|}{a} + C$$

$$\int \frac{dx}{x^2 - a^2} = \frac{1}{2a}\ln\left|\frac{x - a}{x + a}\right| + C$$

Integrals of inverse trig functions

The integrals of inverse trig functions are numerous and complicated. You can find these (and more) easily on the internet or in a reference book such as CRC's *Standard Mathematical Tables*.

Integration by parts

Analogous to the product rule in differentiation, there is a rule for integration called integration by parts. That is, integration by parts is a way to find the integrals of products. The formula is

$$\int u\,dv = uv - \int v\,du$$

where both u and v are functions of x. The "trick" is to factor the radicand and integrating agent into the product of a function and the derivative of another function.

Example: what is the indefinite integral of xe^{3x} ? Let's partition $xe^{3x}dx$ into $u = x$ and $dv = e^{3u}dx$. So we have u and must calculate du, and then must calculate v from dv which is an easy integral in this example.

$$u = x \qquad du = dx$$

$$v = \frac{e^{3x}}{3} \qquad dv = e^{3x}\,dx$$

Now, all need be done is to apply the formula:

$$\int udv = uv - \int vdu$$

$$\int xe^{3x}dx = \frac{xe^{3x}}{3} - \frac{1}{9}\int e^{3x}dx + C$$

Doing the integration,

$$\int xe^{3x}dx = \frac{xe^{3x}}{3} - \frac{e^{3x}}{9} = \frac{3xe^{3x}}{9} - \frac{e^{3x}}{9} = \frac{e^{3x}}{9}(3x-1) + C$$

One must be careful in partitioning *udv*. For example, had we partitioned $xe^{3x}dx$ into $u = e^{3x}$ and $dv = xdx$, we would have

$$u = e^{3x} \qquad du = 3e^{3x}dx$$

$$v = ½x^2 \qquad dv = xdx$$

and applying the formula would yield

$$\int xe^{3x}\,dx = \frac{x^2e^{3x}}{2} - \int \frac{3x^2e^{3x}}{2}\,du$$

which seems more complicated than the original. So if you get "wrapped around the axle" try a different partition!

•

TABLE OF A FEW INTEGRALS

$$\int dx = x + C$$

$$\int x^n\,dx = \frac{x^{n+1}}{n+1} + C \qquad \text{if } n \neq -1$$

$$\int \frac{1}{x}\,dx = \ln|x| + C$$

$$\int \ln x\,dx = x\ln x - x + C$$

$$\int \log_b x\,dx = x\log_b x - x\log_b e + C$$

$$\int e^x\,dx = e^x + C$$

$$\int a^x\,dx = \frac{a^x}{\ln a} + C$$

$$\int \sin x\,dx = -\cos x + C$$

$$\int \cos x\,dx = \sin x + C$$

$$\int \tan x\,dx = -\ln|\cos x| + C$$

$$\int \cot x\,dx = \ln|\sin x| + C$$

$$\int \sec x\,dx = \ln|\sec x + \tan x| + C$$

$$\int \csc x\,dx = -\ln|\csc x + \cot x| + C$$

$$\int \sec^2 x\,dx = \tan x + C$$

$$\int \csc^2 x\,dx = -\cot x + C$$

$$\int \sec x\tan x\,dx = \sec x + C$$

$$\int \csc x\cot x\,dx = -\csc x + C$$

$$\int \sin^2 x\,dx = \frac{1}{2}(x - \sin x\cos x) + C$$

$$\int \cos^2 x\,dx = \frac{1}{2}(x + \sin x\cos x) + C$$

THE
MATH TUTOR'S FRIEND
transform pairs
s-domain
L.T.
$\int_a^b f(x)dx = F(b) - F(a)$
It's easy !!
differential
equations
$\mathfrak{L}[f(t)] = \int_0^\infty f(t)e^{-st}dt = F(s)$
Laplace
transforms
Mr.
Math

3. LAPLACE TRANSFORMS

TABLE OF CONTENTS

page

Introduction to Laplace transforms 3-1
Integro-differential equations
Laplace transforms (L.T.s)

What is a Laplace transform? 3-2
Anatomy and definition of the Laplace transform

Table of Laplace transform pairs 3-3

•

3. LAPLACE TRANSFORMS

Introduction to Laplace transforms

Integro-differential equations

Solving integro-differential equations* is not like solving algebraic equations. The former may contain derivatives and integrals and solutions may not exist or may be unclear and not unique. This is all way beyond the scope of this book which seeks only to introduce Laplace transforms, a method for using algebraic techniques to find solutions to these types of equations. *Integro-differential equations,* containing both derivatives and integrals, are found in many disciplines including engineering, physics, economics, and biology. Here are two examples of *differential equations* (D.E.), a first-order D.E. and a second-order D.E.

$$\frac{du}{dx} = au + x^2$$

$$\frac{d^2u}{dx^2} - x\frac{du}{dx} + u = 0$$

The following is an integro-differential equation for an electrical circuit containing a resistor, inductor and capacitor (V_R, V_L and V_C are voltages across the resistor, inductor, and capacitor, respectively):

$$V_R + V_L + V_C = V_S \Rightarrow iR + L\frac{di}{dt} + \frac{1}{C}\int idt = V_S$$

Laplace transforms (L.T.s)

We learned in algebra that logarithms allow us to convert multiplication of two factors into an addition process. Analogously, Laplace transforms convert integro-differential equations in calculus into algebraic expressions and permit the use of algebra to find solutions. This is done as follows:

1. Transform the differential equation (D.E.) into *Laplace transforms.*
2. Solve the transforms for the unknowns using algebraic methods.
3. Take the inverse Laplace transform to get the answer to the D.E.

The notation "log(*a*)" means "take the logarithm of *a*;" the script "L" notation $\mathcal{L}$ is used to indicate the Laplace transform of a function. For example, "$\mathcal{L}$[f(*t*)]" means "take the Laplace transform of the function f(*t*)." The function variable *t* usually represents time.

* Equations containing only derivatives are differential equations; if only integrals, then *integral equations.*

What is a Laplace transform?

The *Laplace transform* is defined with a new independent variable, *s*, as follows:

$$\mathcal{L}[f(t)] = \int_0^\infty f(t)e^{-st}dt$$

The Laplace transform (*L.T.*) converts a function of time (*t*) into a function of a new variable, *s*, that is independent of *t*. For example, the Laplace transform of cos(ωt) is written as*

$$\mathcal{L}[\cos(\omega t)] = \int_0^\infty \cos(\omega t)e^{-st}dt$$

Anatomy and definition of the Laplace transform

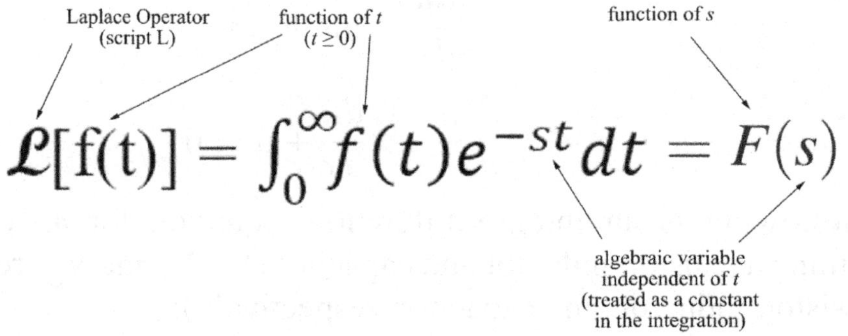

Examples:

$$\mathcal{L}[Ae^{at}] = A\int_0^\infty e^{at}\, e^{-st}\, dt = A\int_0^\infty e^{(a-s)t}\, dt$$

$$= \frac{Ae^{(a-s)t}}{a-s}\bigg|_{t=0}^{\infty} = \frac{A}{(a-s)}\left[e^{(a-s)\cdot\infty} - e^{(a-s)\cdot 0}\right]$$

$$= \frac{A}{(a-s)}[0-1] = \frac{A}{s-a}$$

So, now we have the L.T. of an exponential function. Now what? Well, let's suppose that we have solved a more complicated algebraic expression in *s* and the end result was 10/(*s* – 3). We need only go to a table of *L.T. Pairs* and find

$$Ae^{at} \leftrightarrow \frac{A}{s-a}$$

The solved problem yields the result 10/(s – 3), that is, $A = 10$ and $a = 3$, we can immediately write the solution in the time domain as $10e^{3t}$.

* The "ω" is angular frequency (e.g., degrees/second) and is equivalent to 2πf; that is, ***ω = 2πf***

To summarize, we convert an integro-differential equation from the *time (t) domain* into the *Laplace transform domain (s)*, solve the problem algebraically then transform the result back to the t domain by taking the *inverse L.T.* The table below shows a few *transform pairs*. Comprehensive tables are available in engineering texts and on the internet.

Table of Laplace transform pairs

TRANSFORM PAIRS

$f(t) \quad t \geq 0$	$\mathcal{L}[f(t)] = F(s)$
$A\epsilon^{at}$	$A/(s - a)$
A	A/s
At	A/s^2
At^n	$An!/s^{n+1}$ *
$A \cos \omega t$	$As/(s^2 + \omega^2)$
$A \sin \omega t$	$A\omega/(s^2 + \omega^2)$

*n = positive integer

We stop here. The purpose of this section is merely to alert the math tutor to the existence of Laplace transforms, a technique for solving integro-differential equations in calculus by using high-school level algebra.

●

Index

Symbols

Δx 2-1

A

anti-derivative 2-4, 2-5
antiderivative 2-1

C

chain rule 1-8, 1-12
Chain-sine Rule 1-10
composition of two functions 1-7
concave 1-3
constant of integration 2-1, 2-5
convex 1-3

D

D.E 3-1
definite integral 2-1, 2-2
derivative 1-2
differentiable function 1-2
differential equations 3-1
domain 1-1, 1-2, 1-3, 2-1, 2-2, 3-2, 3-3,
dot notation 1-5

E

extrema 1-3

F

$f \circ g$ 1-7
f(g(x)) 1-7
first derivative 1-3
Fundamental Theorem of Calculus 2-1, 2-5

I

implicit differentiation 1-12
improper integrals 2-3
indefinite integral 2-1
inflection point 1-3
integrable function 2-1
integral equation 3-1
integrating agent 2-5
integro-differential equation 3-1
inverse L.T. 3-3

L

Lagrange notation 1-5
Laplace transform 3-1
Laplace transform domain 3-3
Leibnitz notation 1-5
limit of a function 1-1
limits of integration 2-1
L.T. 3-1, 3-2
L.T. Pairs 3-2

N

Newton's dot notation 1-5

O

operator notation 1-5

P

partial derivative 1-11
partial differentiation 1-6
Power Rule 1-5, 1-6
prime notation 1-5
product rule 1-6

Q

quotient rule 1-6, 1-8

R

Riemann sum 2-3
Riemann Sum 2-1
Rules of Differentiation 1-6

S

second derivative 1-3
second partial derivatives 1-12

T

time (t) domain 3-3
total derivative 1-11
total differential 1-11
transform pairs 3-3

• • •

THE MATH TUTOR'S FRIEND
definite integral
derivatives
f'
It's easy !!
dy/dx
Test Questions
indefinite integrals
Mr. Math
A Look at Calculus

QUIZZES

Acceleration of a dropped ball Q-2
Area problem Q-2

Chain Rule Q-3

Derivative of a function Q-3, Q-4
Domain of a function Q-4

Integration Q-5
Integration by parts Q-6

Laplace Transform Q-6

Time required for ball to drop Q-1

Velocity of a dropped ball Q-1

•

• A ball is dropped from a height of 80 feet. Its height above ground (in feet) t seconds later is given by $s(t) = -16t^2 + 80$. How long will it take for the ball to hit the ground?

A. 2.2 seconds
B. 3.0 seconds
C. 4.1 seconds
D. 5.0 seconds

• • Answer A.
The minus sign indicates a downward velocity The height of the ball will be zero when it hits the ground; that is, $s(t) = -16t^2 + 80 = 0$.
The first term to the right of the " = " sign is the acceleration due to gravity (negative because gravity pulls "downward.") and the second term, the constant 80, is the position of the ball at time $t = 0$. Now solve for t:

$$t^2 = 80/16 \text{ and } t = (\sqrt{80})/4 = \sqrt{(16 \times 5)}/4 = \sqrt{5} \approx 2.2 \text{ seconds}$$

• ♥ •

• A ball is dropped from a height of 80 feet. Its height above ground (in feet) t seconds later is given by $s(t) = -16t^2 + 80$. What is the velocity of the ball when it hits the ground?

A. 71.6 ft/sec
B. -71.6 ft/sec
C. 66.7 ft/sec
D. -66.7 ft/sec

• • Answer B. The minus sign indicates a downward velocity (Answer A would be correct if the speed instead of velocity were requested).
The height of the ball will be zero when it hits the ground; that is,

$$s(t) = -16t^2 + 80 = 0.$$

The first term to the right of the " = " sign is the acceleration due to gravity (negative because gravity pulls "down.") and the second term, the constant 80, is the position of the ball at time $t = 0$. Now solve for t:

$$t^2 = 80/16 \text{ and } t = (\sqrt{80})/4 = \sqrt{(16 \times 5)}/4 = \sqrt{5} \approx 2.2 \text{ seconds}$$

The velocity is the first derivative of s(t): $s'(t) = -32t$, and so at $f = \sqrt{5}$ seconds the velocity will be $-32 \times \sqrt{5} = -71.6$ ft/sec, again, negative because the movement is downward.

• A ball is dropped from a height of 80 feet. Its height above ground (in feet) t seconds later is given by $s(t) = -16t^2 + 80$. What is the acceleration of the ball when it hits the ground?

A. 32.2 ft/sec^2
B. -32.2 ft/sec^2
C. 65.4 ft/sec^2
D. -64.4 ft/sec^2

• • Answer B. The minus sign indicates a downward velocity The height of the ball will be zero when it hits the ground; that is, $s(t) = 0$; that is, $s(t) = -16t^2 + 80 = 0$. The first term to the right of the "=" sign is the acceleration due to gravity (negative because gravity pulls "down.") and the second term, the constant 80, is the position of the ball at time $t = 0$. Now solve for t:

$$t^2 = 80/16 \text{ and } t = (\sqrt{80})/4 = \sqrt{(16 \times 5)}/4 = \sqrt{5} \approx 2.2 \text{ seconds}$$

The velocity is the first derivative of s(t): $s'(t) = -32t$, and so at $f = \sqrt{5}$ seconds the velocity will be $-32 \times \sqrt{5} = -71.6$ ft/sec, again, negative because the movement is downward.

What is the acceleration of the ball when it hits the ground? The acceleration is the second derivative of s(t): $s''(t) = s'(s'(t)) = -32$ ft/sec ^{2}, as expected because this is the acceleration due only to gravity (i.e., there was no initial acceleration by, say, throwing the ball down instead of letting it fall). The gravitational constant for Earth

• ♥ •

is -32.2 ft/sec ^{2}, approximately.

• A rectangle has length three times its width. What is the maximum area that can be enclosed by a perimeter of 250 ft?

A. 1150 square feet
B. 2125 square feet
C. 11719 square feet
D. 9723 square feet

• • Answer C.

Let y and x be length and width.

Given: $y = 3x$ and $2x + 2y = 250$ from which $y = 125 - x$

$A = \text{area} = xy = x(125 - x) = 125x - x^2$

$A' = 125 - 2x$ from which the critical point is $x = 62.5$

$A'' = -2$, so the critical point is a maximum

The dimensions for maximum area are $x = 62.5$ ft and $y = 3x = 3\times62.5 = 187.5$ ft and the area is $62.5\times187.5 \approx 11\,719$ square feet, a little more than a quarter of an acre.

• The derivative of $y = 2\sin(2x^2 + 3x)$ is [Hint: apply the chain rule]

A. $2\cos(2x^2 + 3x)$
B. $-2\cos(2x^2 + 3x)$
C. $8x\cos(2x^2 + 3x)$
D. $2\cos(2x^2 + 3x)(4x + 3)$
E. None of the above

• • Answer D.
Let $u = (2x^2 + 3x)$. $du = 2x + 3$

$$\frac{dy}{dx} = 2\cos u \cdot \frac{du}{dx} = 2\cos(2x^2 + 3x) \cdot (4x + 3)$$

• ♥ •

• The derivative of e^x with respect to x is

A. e^x
B. $2e^x$
C. $e^x/2$
D. e^{2x}
E. None of the above.

• • Answer A.

• Find the derivative of $y = e^{2\sin(2x^2)}$

A. $e^{2\sin(2x^2)}$

B. $e^{2\cos(2x^2)}$

C. $8xe^{2\sin(2x^2)} \cdot \cos(2x^2)$

D. 0

E. None of the above

• • Answer C.

Example: Find the derivative of $y = e^{2\sin(2x^2)}$¶

Using the Chain Rule, let $u = 2\sin(2x^2)$, then $\frac{du}{dx} = 2\cos(2x^2)\cdot 4x$¶

so, $y = e^u$ and $\frac{dy}{dx} = e^u \frac{du}{dx} = e^{2\sin(2x^2)} \cdot 8x\cos(2x^2) = 8xe^{2\sin(2x^2)} \cdot \cos(2x^2)$¶

• ♥ •

• The domain of $f(x) = 1/x$ is

A. $(-\infty, \infty)$
B. $(0, \infty)$
C. $(-\infty, 0) \cup (0, \infty)$
D. $(-\infty, 0) \cap (0, \infty)$
E. None of the above

• • Answer C.

- $\int_0^\infty 3e^{-x} =$

A. 0
B. 3
C. 6
D. ∞
E. -∞

• • Answer B.

$$\int_0^\infty 3e^{-x} = -3(e^{-\infty} - e^{-0}) = -3(0 - 1) = 3$$

• ♥ •

• Perform the integration:

$$\int 4x^3 dx$$

A. x^4
B. $x^4 + 247$
C. $x^4 - 247$
D. All of the above
E. None of the above

• • Answer D. The answer is $x^4 + C$, where C is a constant. The derivative of any of the answers shown will be $4x^3$, the integrand.

- What is the solution to the following integral?

$$\int xe^{3x}dx$$

A. 3
B. Undefined
C. $\pm\infty$
D. a function
E. None of the above

• • Answer D.

Partition $xe^{3x}dx$ into $u = x$ and $dv = e^{3x}dx$. So we have u and must calculate du, and then must calculate v from dv which is an easy integral in this example.

$$u = x \qquad du = dx$$

$$v = \frac{e^{3x}}{3} \qquad dv = e^{3x}\,dx$$

Now, all need be done is to apply the formula:

$$\int u\,dv = uv - \int v\,du$$

$$\int xe^{3x}dx = \frac{xe^{3x}}{3} - \frac{1}{3}\int e^{3x}\,dx$$

Doing the integration,

$$\int xe^{3x}dx = \frac{xe^{3x}}{3} - \frac{e^{3x}}{9} = \frac{3xe^{3x}}{9} - \frac{e^{3x}}{9} = \frac{e^{3x}}{9}(3x-1) + C$$

which is a function. Note that there are an infinity of solutions because C can take on any value.

• ♥ •

- L[Ae^{at}] =

A. A(s – a)
B. (s – a)/A
C. A/(s – a)
D. Ae^{-as}/(s – a)
E. None of the above

$\mathcal{L}[Ae^{at}] = A\int_0^{\infty} e^{at}\,e^{-st}\,dt = A\int_0^{\infty} e^{(a-s)t}\,dt$¶ • • Answer C.

$$= \frac{Ae^{(a-s)t}}{a-s}\bigg|_{t=0}^{\infty} = \frac{A}{(a-s)}\left[e^{(a-s)\cdot\infty} - e^{(a-s)\cdot 0}\right]$$¶

$$= \frac{A}{(a-s)}[0-1] = \frac{A}{s-a}$$¶

• The slope of a line tangent to f(x) may be expressed as

A. $f'(x)$
B. $\frac{dy}{dx}$
C. $D_xF(x)$
D. Any of the above
E. None of the above

• • Answer D.

• ♥ •

• Perform the integration:

$$\int_2^3 4x^3\, dx =$$

A. 16
B. 65
C. 81
D. $12x^3$
E. None of the above

• • Answer B.

$$\int_2^3 4x^3\, dx = 3^4 - 2^4 = 81 - 16 = 65.$$

Note that the indefinite integral produces a function (which, when differentiated yields the integrand) and the definite integral produces a number.

THE MATH TUTOR'S FRIEND

INFORMAL REVIEW OF HIGH-SCHOOL MATH TOPICS FOR TUTORS

- 8 -

• A Look at Calculus •

Made in the USA
Monee, IL
05 November 2026

67386651R00033